PAST & PRESENT

SOUTHWESTERN PENNSYLVANIA'S COAL REGION

WESTMORELAND COAL COMPANY BUILDING, EXPORT. The Westmoreland Coal Company Building seen in this photograph was in the town of Export in Westmoreland County, Pennsylvania. (Courtesy of Westmoreland County Historical Society.)

Past & Present

Southwestern Pennsylvania's Coal Region

Marion M. Piccolomini

ISBN 978-1-4671-0417-3

Library of Congress Control Number: 2019947030

Published by Arcadia Publishing
Charleston, South Carolina

Printed in the United States of America

For all general information, please contact Arcadia Publishing:
Telephone 843-853-2070
Fax 843-853-0044
E-mail sales@arcadiapublishing.com

Visit us on the Internet at www.arcadiapublishing.com

On the Front Cover: Westmoreland Coal Company South No. 2 Mine Entrance, Export. The Westmoreland Coal Company opened the Export mine in 1892, and by the early 1900s, there were two drift mine entrances: Export No. 2 (South No. 2 Mine, seen here) and Export No. 1 (North Mine). In the past image, cars loaded with coal are exiting the mine. Through the efforts of the Export Historical Society, the entry on the right in the present image has been cleared, and work is being done to uncover the left side. See page 38. (Top, courtesy of Export Historical Society; bottom, author's collection.)

On the Back Cover: Winter Scene of Beehive Coke Ovens. The snow provides a dramatic background for this scene outlining the structure of the ovens and highlighting the lorry/larry car (center) and women on top of the row of coke ovens. (Courtesy of Uniontown Public Library.)

Contents

Acknowledgments

This book was possible because of the help and support of many individuals, historical societies, and organizations that shared photographs, artifacts, historical information, and personal stories that embrace the rich history of the accomplishments of the industries, communities, and people of the region. Thank you to the following and others I spoke to who I may have missed: Archives & Special Collections, University of Pittsburgh Library System; Bill Argenti; Bill Bryner; California Area Historical Society; Coal and Coke Heritage Center at Penn State Fayette, the Eberly Campus; Dan Hiteshew; Detre Library & Archives, Heinz History Center; Donald Lancaster; Dunbar Historical Society, with special thanks to Donna R. Myers; Ellen R. Keder; *Engineering and Mining Journal* (E&MJ); Eureka Coal Heritage Foundation, with special thanks to Matt Scislowicz; Evan T. Williams II; Export Historical Society, with special thanks to Melanie Litz and Mike Mance; Fayette County Historical Society, with special thanks to Chris Buckelew; Fiore Piccolomini; Fredericktown Public Library; George Milne; Historical & Genealogical Society of Somerset County; Hagley Museum and Library; Jim Shorkey Uniontown Kia; Mike and Kathy Packrall; Marianna Community Public Library; Mount Pleasant Historical Society, with special thanks to Gene Kowalewski, Bill Hare, and Rick Meason; Pennsylvania State Archives; *Pittsburgh Post-Gazette*; Pittsburgh History & Landmarks Foundation; Thomas E. Rusnack; Rivers of Steel National Heritage Area, with special thanks to director of historic resources and facilities Ronald Baraff; Sisters of the Order of St. Basil the Great; Scottdale Historical Society Inc., with special thanks to Thomas Zwierzelewski;Special Collections and University Archives of Indiana University of Pennsylvania, with special thanks to Dr. Harrison Wick; the Frick Collection/Frick Art Reference Library Archives; the Frick Pittsburgh; Uniontown Public Library; Westmoreland County Historical Society; and West Overton Village and Museum Archives, with special thanks to Aaron Hollis Jr.

Special thanks to my husband, John, and son Mark for being a part of the photographic journey and for the help through the ongoing review process. Your help was invaluable.

To my sister Carol for her help with the review process and who spent endless hours traveling the region, many times to parts unknown. Being able to share the experience has been a treasured part of the book.

Special thanks to Amanda Peters, archivist at the Coal and Coke Heritage Center at Penn State Fayette, the Eberly Campus, whom I met at the center on one of my first fact-finding missions. Her knowledge of the museum's collection provided not only access to the relevant history stored there, but also insights into the background of the collection. Thank you as well for your assistance in the review of the material and images in this book.

Finally, to my nephew, who helped me see the vision for this book as we toured the region to areas that were just names of places until he shared the history of the past: thank you for providing professional knowledge of the coal and coke industries and access to your extensive library of the history of the region. Without your subject matter expertise, technical insight, and tireless editing, this book would not have been possible.

Introduction

Coal is an important part of the history of southwestern Pennsylvania. Early in its history, the mostly rural region became known as the "gateway to the west" as people were drawn to the wilderness beyond the Allegheny mountains. When the United States began to move to greater industrialization in the 1800s, the natural resources of the region played a valuable role in the development of the area. There were waterways for transportation and coal and wood to fuel the growth of industry and the communities these new industries would bring. In western Pennsylvania, the expansion of highways, railways, and waterways made the area more accessible to outside markets, which was especially important for the bituminous coal region located mainly in rural areas. Roadways expanded and improved, and railroads frequently built spur lines or connections to coal and coke operations, which made it easier to ship coal beyond the coal mining regions. Early in the industry, individuals or small operations mined coal from hillsides for local use. As industrial demands for coal grew, so did the mining industry, and larger coal mining companies developed in the region to meet domestic and industrial needs. Of the two types of mining—surface and underground—most operations during the coal era in southwestern Pennsylvania were deep or underground mines. The three types of underground mining—drift, slope, and shaft—were found throughout the region.

As the production of steel became a dominant industry, the demand for coal grew. Coke (a carbon-rich product made from coal) was one of the crucial materials used to make the molten iron necessary to produce steel. Coal and coke became a vital part of the steel production cycle. Of special importance to the development of the region was the high-quality coal found in the Pittsburgh Coal Seam. The coal from the Connellsville section of the Pittsburgh Seam with a high carbon content and a low content of sulfur and other impurities made it excellent for coke production. According to *Coal and Coke Resource Analysis Western Pennsylvania, Northern West Virginia*, coking became the dominant industry in the area, since it had the highest-quality coal for making coke at the time.

The relationship between the coal, coke, and steel industries resulted in an economic boom for the area, and the stage was set for southwestern Pennsylvania to play an important role in the industrial development of the United States, and it led in coal production from the end of the 19th century into the mid-20th century. Coal mines, coke ovens, and steel mills were common sights around the region as parts of the area experienced a transition from a rural, agriculture-based economy, to an industrial base. People from around the country along with a great number of European immigrants were drawn to the region for work. Communities expanded and new towns were built by mining companies as a large influx of people came to the area. Like other businesses at the time, management and labor had to deal with the issues of the new era of industrialization, including production demands, workplace safety, labor movements, and environmental issues. The coal industry evolved to meet these challenges. In the early to mid-20th century, the industry saw a decline. Some of the contributing factors leading to it were reduction in the demand for steel, increased efficiency in the steel-making process, use of other fuels, depleted coal reserves, introduction of mechanized mining, and the switch from beehive to by-product coke ovens built near steel mills. Just as the interrelationship between coal, coke, and steel led to their growth, it also led to their decline, and many coal and steel operations in the region closed toward the end of the 20th century. Although production is nowhere near the early

20th century, coal and steel production continues in the region, and other industries and businesses have moved into the area.

I was lucky enough to have grown up in a small town in the bituminous coal region of southwestern Pennsylvania where neighbors spoke in foreign languages when they did not want the children to understand what they were saying, cabbage rolls and Italian pork sandwiches were familiar foods at picnics and social gatherings, and the whole town looked forward to festivals at local churches or town centers. Everyone knew everyone, and the families were representative of the melting pot of the time. I just assumed that every town had a Catholic church, a Baptist church, a Polish club, and a Sons of Italy social club and that everyone had neighbors who proudly shared their cultural heritage—usually through food. The town, then and now, is surrounded by communities of varying sizes, some of which are so close in proximity that it is difficult to tell where one ends and the next begins. When I was growing up, most of the coal, coke, and steel industry were declining. Even though surrounded by reminders of the industries that once boomed in the area, it was easy to take for granted the story of how the hard work, dedication, and struggles of the past led to better working conditions and the lifestyle we have today.

This book shares the importance of the region and the uniqueness of the area across several counties in southwestern Pennsylvania through pictures. Images of sites when the industries dominated the region are paired with pictures of those sites today, integrating the past with the present and paying tribute to the people and industries that made it all work, setting the foundation for the region today.

CHAPTER 1

COAL-COKE-STEEL

FILBERT MINE AND HOMES (BACKGROUND). The Filbert Mine, in Fayette County, opened by the H.C. Frick Coke Company, was one of many operations in the region. The buildings seen here provided support for the underground operation. The headframe (tall structure at center) brought the coal to the surface and to the tipple (left of the headframe), where it was processed. How the coal was processed depended on the operation. Coal from some area mines was shipped directly to customers. Other coal, especially from the Connellsville district, was coked locally and then shipped to customers like the steel mills in the Pittsburgh area. The homes on the hillside were built and owned by the company for workers. The company store (at right, across from the homes), was within walking distance from the town. Businesses and services moved into the area to support the development of these industries and the communities that supported them, changing the landscape and culture of the region. This chapter includes images from Allegheny, Fayette, Greene, Indiana, Somerset, Washington, and Westmoreland Counties. (Courtesy of Pennsylvania State Archives, MG-218.1 General Photograph Collection, Industry-Mines, Cramer Collection, H.C. Frick Co. Photographs, Image 3001.)

View of Fort Duquesne and the Confluence of the Rivers in 1758 and Aerial View of Pittsburgh with Stadiums. According to the "First Mining of Pittsburgh Coal" historical marker in Pittsburgh, the state's bituminous coal industry was born around 1760 on Coal Hill, now Mt. Washington (below, right). Pittsburgh was positioned to become a top producer of steel with a nearby supply of coal for fuel and access to railroads and waterways for transportation. Today, as seen above, this area is surrounded by urban development. (Below image courtesy of AIS Postcard Collection, Archives & Special Collections, University of Pittsburgh Library System. Above image courtesy of Copyright ©, Pittsburgh Post-Gazette, 2019, all rights reserved. Reprinted with permission.)

Homestead Steel Works. 42-Inch Plate Mill. Andrew Carnegie purchased the Homestead Works on the Monongahela River in the early 1880s. He sold his interest in Carnegie Steel Corporation, and the Homestead Mill became part of the newly formed US Steel Corporation. It was a dominant force in the industry at that time. Homestead closed in the mid-1980s. The working mill has been transformed, and visitors can now find shops, restaurants, hotels, and entertainment at the Waterfront Shopping Center. (Above, courtesy of William J. Gaughan Collection, AIS.1994.03, Archives & Special Collections, University of Pittsburgh Library System; below, courtesy of the Waterfront.)

Carrie Furnace. Carrie Furnace was purchased by Andrew Carnegie. Over time, additional furnaces were added. The furnaces produced molten iron for the Homestead Steel Works. The iron was transported across the Rankin Hot Metal Bridge, which was designed to transport iron in special railroad cars so that there was no spillage into the river. (Above, courtesy of William J. Gaughan Collection, AIS.1994.03. Archives & Special Collections, University of Pittsburgh Library System; below, author's collection.)

Carrie Furnace. Furnaces 6 and 7 were built in 1906–1907. They opened in 1907 and were taken off-line in 1978. They provided iron for the Homestead Mill. They are the only furnaces still standing from the operation. The two furnaces were designated a National Historic Landmark in 2006, and the Rivers of Steel National Heritage Area provides guided tours of the furnaces. (Above, courtesy of John Viglione; below, courtesy of Rivers of Steel National Heritage Area.)

River Car Ferry and Barges. Jones and Laughlin Steel Corporation (J&L) was another force in the Pittsburgh steel industry. Its Pittsburgh Works/Southside Works (below) was along the Monongahela River in the area now known as the South Side. Iron produced at the Eliza blast furnaces across the river was transported over the Hot Metal Bridge to the mill. J&L was bought by LTV Steel in 1968. It eventually closed, and Eliza Furnace was demolished in 1983. Southside Works (above), a retail, office, and residential complex, is built on a portion of the old mill site. (Above, courtesy of anonymous; below, courtesy of Jones & Laughlin Steel Corp. Photographs, Detre Library & Archives, Heinz History Center.)

Banning No. 1, 1933 and 2017. Banning No. 1 was near the Fayette County–Westmoreland County border along the Youghiogheny River. The concrete cylinder near the railroad siding and tipple at right in the 1933 image is still standing. A sign nearby along the Great Allegheny Passage rail-trail describes the site as a large coal processing operation of the Pittsburgh Coal Company. (Above, courtesy of Pittsburgh and Lake Erie Railroad Company Photographs, 1850-1999, AIS.1982.23, Archives & Special Collections, University of Pittsburgh Library System; below, author's collection.)

Coke Yard, Leisenring No. 2. The photograph below shows a portion of the site of the Leisenring No. 2 Mine in Bute/West Leisenring. This operation was near the Fayette County seat of Uniontown. It was one of three Leisenring operations. The mine closed in the 1950s. Some of the coke ovens pictured above remain (see chapter two). (Above, courtesy of Rev. Edward C. Tajc Photographs, Patch Work Voices Community Collection, Coal and Coke Heritage Center at Penn State Fayette, the Eberly Campus; below, author's collection.)

Oliver No. 1 Shaft. There were three Oliver mines, known as Oliver 1, 2, and 3. The headframe (tallest structure seen below) was located above the vertical shaft that went to the coal seam below. Coal was hoisted out of the mine and processed for coking or for shipment to customers. The Oliver operations are closed. The brick building at the previous site of Oliver 1 today (above), is not visible in the past image from the *Engineering Magazine* XX, 1900, but remains a reminder of the history of the area. (Above, author's collection; below, courtesy of Hagley Museum and Library.)

Mine Shaft at One of the Oliver Mines. The hoist house (below, left) is connected to the headframe, which is also seen in the artist's rendition of the Oliver Mines above. The painting is a combination of the signature elements of the multiple phases of a coal and coke operation in the region. The coal was brought out of the mine and transported by rails to the coke ovens for coking. The uniform houses of the company town can be seen in the distance on the hillside. (Above, courtesy of Fiore Piccolomini; below, courtesy of Patch Work Voices Community Collection, Coal and Coke Heritage Center at Penn State Fayette, the Eberly Campus.)

Phillips Mine Buildings. The Phillips Mine and Coke Works was a few miles from Oliver and West Leisenring. According to *Patches of History: The 1920's & 1930's Heyday of Fayette County Coal & Coke in Pennsylvania*, "it was named for the Phillips family who owned the land." Below is the general site of the operation today, behind a business on Route 51. (Above, courtesy of Pennsylvania State Archives, MG-218.1 General Photograph Colleciton, Industry-Mines, Cramer Collection, H.C. Frick Co. Photographs, Image 3002; below, author's collection.)

Memorial Service at Star Junction. About 12 miles north of Phillips is the town of Star Junction. According to *A Legacy of Coal: The Coal Company Towns of Southwestern Pennsylvania*, James Cochran of Dawson was known as one of the greatest pioneers of the coke industry. He founded the Washington Coal and Coke Company, which operated the plant at Star Junction. Two buildings that remain from the operation can be identified in both images: the redbrick building at right and the one-story building at left. (Above, courtesy of the Fayette County Historical Society; below, author's collection.)

Bridge, 1930 and 2014. Railroad lines expanded into the bituminous coal region, including the Mather Mine in Greene County, linking once isolated areas to outside markets. The image below from April 1930 shows a bridge on the Mather Colliery Company siding from the bank of Tenmile Creek. The remains of the bridge are seen above in 2014. (Above, courtesy of Evan T. Williams II; below, courtesy of Monongahela Railway Company Photographs, 1903-1993, AIS.2009.07, Archives & Special Collections, University of Pittsburgh Library System.)

COAL TRAIN. The coal train below has just crossed a bridge in Poland, Pennsylvania, in 1941. The Poland Mine in Greene County is no longer in operation. Coal is still transported by train however, as seen above in the image of a train transporting coal today in Fayette County. (Above, courtesy of John Viglione; below, courtesy Monongahela Railway Company Photographs, 1903-1993, AIS.2009.07, Archives & Special Collections, University of Pittsburgh Library System.)

Clearfield Bituminous Coal Company, Office Building, Water Street, Indiana, 1920 and 2018. According to "Clearfield Bituminous Coal Corporation land tract records and property maps, circa 1885–2009," the company purchased land in several counties, including Indiana, and by 1919, its mines there produced three-quarters of the corporation's coal. The building looks much as it did in 1920. Today, there are county offices inside. (Above, courtesy of Rochester and Pittsburgh Coal Company Media Collection, Special Collections & University Archives, Indiana University of Pennsylvania; below, author's collection.)

Rochester and Pittsburgh Coal Company Office Building, Indiana, 1970 and 2018. The Rochester and Pittsburgh Coal Company opened mines in multiple counties, including Indiana County. Though the building is occupied by other tenants today, the company's name, Rochester and Pittsburgh Coal Company, can still be seen (above the awning below). (Above, courtesy of Rochester and Pittsburgh Coal Company Media Collection, Special Collections & University Archives, Indiana University of Pennsylvania; below, author's collection.)

Berwind White Coal Mining Company Office, Windber. The company's office building in Windber, built in 1903, looks remarkably similar today. The Berwind White Coal Mining Company had mines in several counties in Pennsylvania when it began to purchase property in Somerset County in the 1890s. In addition to its office building, the company helped finance other amenities in the town (as seen in chapter three). This was the municipal building for a time, and today, it remains a reminder of the era. (Above, author's collection; below, courtesy of Collection of the Historical & Genealogical Society of Somerset County.)

Coal Train Leaving Plant, 1920s and 2019. The coal train seen below is passing through Windber in Somerset County. The building at left and the main Eureka Department Store building (right) can be seen in the background. (Above, author's collection; below, courtesy of Eureka Coal Heritage Foundation.)

Phillips Mine and Billy Lowe. In early, pre-mechanized mines, picks and shovels were used to dig coal with open-flame lights providing the only source of illumination. Coal was hand loaded, and before regulations prohibited it, boys sometimes worked with their fathers to increase the amount of coal mined since miners were paid by the ton. It was dangerous and difficult work, and there were frequent accidents. The use of machines like the one below became more common toward the 1950s. (Above, courtesy of US Steel Corporation Records, Coal and Coke Heritage Center at Penn State Fayette, the Eberly Campus; below, courtesy of Rochester and Pittsburgh Coal Company Media Collection, Special Collections & University Archives, Indiana University of Pennsylvania.)

PANORAMIC VIEW OF RACHEL MINE, TOWN OF MARIANA IN THE BACKGROUND 1911. The Rachel Mine was one of the shafts of the Marianna Mine, owned by the Pittsburg-Buffalo Company. According to *A Bicentennial History of West Bethlehem Township and Marianna Borough 1776–1976*, Pres. Theodore Roosevelt invited mining experts from Europe to inspect the coal industries of the United States, which included a tour of the Marianna Plant in October 1908. Forty-five days later, a disaster occurred at the mine (see chapter five). Both images show the homes built by the company on the hillside and the remains of coke ovens below at lower right. (Above, courtesy of *Coal, Coke, Brick, Sewer Pipe and Builders' Supplies*. Historic Pittsburgh Book Collection, Archives and Special Collections, University of Pittsburgh Library System; below, author's collection.)

VESTA NO. 4 MINE. Vesta No. 4 Mine was owned and operated by the coal division of Jones and Laughlin Steel Corporation. In addition to the main portal or entrance, there were other portals, including ones in places like Daisytown and Richeyville. The keystone marker above is at the California Area Historical Society in California, Pennsylvania. According to the historical society, there were "keystones at entrance of Vesta No. 4 Mines. California, Richeyville, Smallwood, and Vestaburg each had a stone." (Above, photograph by the author with the permission of the California Area Historical Society; left, courtesy of CONSOL Energy Inc. Mine Maps and Records, 1857–2010, AIS.1991.16, Archives and Special Collections, University of Pittsburgh Library System.)

Vesta No. 4 Mine in Richeyville During Peak Production and Mine Buildings after Renovation, 2002. The large brick building on the right below is the same building above at left after being renovated in 2002 by a privately owned business. The Vesta mines were "captive" mines, which means that all the coal from the mines was used by Jones and Laughlin, which owned them. The company's barges traveled on the Monongahela River from the Vesta Mines to its steel operations in Pittsburgh. (Both, courtesy of Mike and Kathy Packrall.)

Bathhouse Richeyville and Vesta No 4 Mine Buildings after Renovation 2002. Some mining operations had a bathhouse, or wash house, where employees could shower at the end of their shift. The layout of the bathhouse depended on the mining company. Note the height of the windows in both images. The bathhouse was renovated in 2002 and is across from the lamphouse, which has also been renovated. (Above, courtesy of Fredericktown Public Library; below, courtesy of Mike and Kathy Packrall.)

VESTA BRIDGE. The support pillar at left in both images is part of the bridge built by the company to transport coal from Vesta 4 and 5 in Washington County to the LaBelle Processing plant across the river in Fayette County. According to the *Pittsburgh Post-Gazette* article "A Past, but No Future: Suspension Bridge Coming Down" on February 29, 2004, "After the war, the Vesta Coal Co. began construction of the Labelle prep plant." (Above, courtesy of Fredericktown Public Library; below, courtesy of John Viglione.)

VESTA BRIDGE ON ROUTE 88, 2018 AND 2003. LTV bought J&L in the late 1960s and continued to operate the processing plant until it was sold in the early 1980s. After transporting tons of coal across the river, the Vesta Bridge was set to be demolished. Although it was not being used at the time of the 2003 photograph below, large portions of the structure were still present. Today, little remains of the bridge. (Above, author's collection; below, courtesy of Jonathan Tajc.)

Miner's hat with Sunshine Lamp, 1890s–1920s, and Carbide Lamp. According to the Fayette County Historical Society, the above "used Naphtha as a fuel with an open flame and were more dangerous than the later carbide lamps." The open flame was a danger since it could ignite explosive gases present in the mine. The carbide lamp below was less dangerous, but still had a flame. The push for improved safety and working conditions eventually led to safe, battery-powered lights used today. (Above, courtesy of the Fayette County Historical Society; below, courtesy of Bill Argenti.)

Birdcage for Canary to Detect Methane, 1920s–1940s, and Flame Safety Lamp. According to the Fayette County Historical Society, "Since methane is odorless the canary was used as a sentinel since they were more sensitive to small amounts of the methane." Small and large mining operations recognized the importance of monitoring air quality. As the industry advanced, systems like the flame safety lamp at left were developed. The flame in these lamps would increase in size when more methane was present. Today, electronic handheld monitoring devices have replaced flame safety lamps and are used to monitor the quality of the air and to provide an alert system for the presence of dangerous gases. (Above, courtesy of the Fayette County Historical Society; left, courtesy of Jonathan Tajc.)

WESTMORELAND COAL COMPANY SOUTH (NO. 2) MINE ENTRANCE (ABOVE) AND EXPOSED SOUTH MINE (NO. 2). Through the restoration efforts of the Export Historical Society, the right portal in both images has been cleared. The image below shows the work being done to restore the left side. The project includes adding signage with historical information and building a sidewalk from the entrance to the uncovered remains of one of the fan houses, a short walk to the left. There, visitors can see the remains of one of the two fan houses that provided circulation to the mine. (Both, courtesy of Export Historical Society.)

MAMMOTH COKE WORKS. The photograph below from a 1900 issue of *The Engineering Magazine* provides a view of the operation with mine buildings (front) and coke ovens (background, left). The building seen above, which is the Mount Pleasant Township Municipal Authority Building, was the Frick lamphouse for the mine, according to a 2002 article in the *Tribune-Review*. The building has not been clearly identified in the picture below. Further information about the mine can be found in chapter five. (Above, author's collection; below, courtesy of Hagley Museum and Library.)

STANDARD SHAFT TIPPLE DATE UNKNOWN. In the late 1870s, A.A. Hutchinson and Brothers Company opened Standard No. 1. H.C. Frick Coke Company bought the operation, and under his ownership, the mine became the largest coal and coke operation in the world, with over 900 beehive ovens. The mine produced tons of coal until the mine and coke works closed in 1931. Today, the one-story building seen at left remains. (Both, courtesy of Gene Kowalewski.)

CHAPTER 2

Coke Production

Winter Scene of Beehive Coke Ovens. This serene image of finely dressed women atop a snow-covered bank of coke ovens is a stark contrast to a working coke operation. Coke is produced by baking or heating coal in a sealed chamber at high temperatures to burn away impurities. It was porous, light, and easier to ship than the soft bituminous coal from the Pittsburgh Seam. So, many mine operations built coke ovens near the mines. Different types of ovens, like the rectangular and byproduct ovens, could be found in the region. But the beehive oven (probably named for its dome-like shape) was the most prevalent in the area. As the steel industry grew, so did the coke industry, especially in the Connellsville coke region that was known for quality coal for coking. As time passed, advances in technology, a declining marketplace, and the increasing awareness of the negative environmental impact of the ovens led to the use of by-product ovens that were often built closer to steel mills. The coal regions in general—and, specifically, the Connellsville region—felt the effects of these changing dynamics. (Courtesy of Uniontown Public Library.)

Morewood Mine and Coke Works Coal Tipple and Mine Buildings. The large white superintendent's house seen in the left background of both images is visible today from Route 119 near Scottdale. In the Connellsville region, many mining operations also had coke ovens (above, right). Coal was loaded into the ovens for coking, and then the coke was loaded for shipment to customers. Mining operations that mined coal for coking but did not have local coke ovens transported the coal to other sites or shipped the coal directly to customers for coking. (Above, courtesy of West Overton Village and Museum Archives; below, author's collection.)

Coke Ovens and Car, Leisenring No. 2, c. 1950 and 2018. The image below shows coke ovens that were part of Leisenring No. 2 in West Leisenring. The lorry/larry car on top of the ovens was used to drop coal into the ovens. The front of the oven was sealed, with only a small opening for airflow into the oven. The coal was usually baked for 48 hours (72 hours on the weekend), and the temperature in the interior averaged 2,500 degrees according to "Coke Making." Although it is not the exact same set of ovens, the image above shows the remains of some coke ovens today. (Above, author's collection; below, courtesy of Rev. Edward C. Tajc Photographs, Patch Work Voices Community Collection, Coal and Coke Heritage Center at Penn State Fayette, The Eberly Campus.)

Coke Ovens. The image above shows abandoned coke ovens that were part of Leisenring No. 2 mine in West Leisenring. The dormant ovens are in stark contrast to the glow from the Leisenring ovens in operation seen below. After the baking process, the oven was unsealed, and workers sprayed water to cool the coke. The coke was then drawn from the oven and loaded for shipment. (Above, author's collection; below, courtesy of Rev. Edward C. Tajc Photographs, Patch Work Voices Community Collection, Coal and Coke Heritage Center at Penn State Fayette, The Eberly Campus.)

MEN WORKING AT COKE OVENS. Similarities can be seen in the structures of the working ovens above and the 2018 photograph below of coke ovens that were part of the Carpentertown operation in Westmoreland County. Early in the industry, coke was drawn by hand. (Above, courtesy of Fayette County Historical Society; below, author's collection.)

Phillips Coke Plant, 1927 and 2018. The coke ovens that lit the sky at Phillips were not far from West Leisenring. Today, the brick building along Route 51 is believed to be the remains of the mechanic shop from the Phillips operation. The coke ovens would have been in the distance, behind the building and to the right. (Above, courtesy of H.C. Frick negative print, U.S. Steel Corporation Records, Coal and Coke Heritage Center at Penn State Fayette, the Eberly Campus; below, author's collection.)

Tipple, Washer, Coke Ovens, and Power Plant At Rachel Mine, Marianna PA. The Marianna Mine also had coke ovens. The ovens were constructed with bricks, like the ones used to build the company houses. Today, remains of the brick coke ovens can be seen from a road in town. (Above, courtesy of the John Viglione collection; below, courtesy of *Coal, Coke, Brick, Sewer Pipe and Builder's Supplies.* Historic Book Collection, Archives & Special Collections, University of Pittsburgh Library System.)

Carpentertown Mine Coke Oven Workers. Carpentertown Mine was one of the few operations to have both beehive and rectangular ovens. The rectangular ovens had metal doors that opened and closed on both sides. Note the closed doors behind the workers below. Similar to beehive ovens, the coal was dropped into the ovens from carts that ran on rails on top of the ovens. There are no rectangular ovens remaining at the site today. (Above, author's collection; below, courtesy of Bill Hare.)

Carpentertown Mine Coke Pusher Machine. The two men above are standing in front of the rectangular coke ovens next to a coke pusher machine. The metal doors opened on both sides of the oven, and the machine was used to push the coke out to the opposite side of the oven after baking. Rectangular ovens were more expensive to build, but cut labor costs. There are no remains of the rectangular ovens at the site of the old Carpentertown mine today. (Above, courtesy of Bill Hare; below, author's collection.)

Carpentertown Mine Coke Receiver/Loader Machine. The receiver/loader machine was used to remove the coke from the rectangular ovens. Coke was pushed out from the other side of the oven to the loader using the coke pusher machine. The coke was then loaded into railroad cars or other vehicles for shipment. Although there are no remains of the rectangular ovens, they were probably to the left of the remains of the beehive ovens still at the site today (below). (Right, courtesy of Bill Hare; below, author's collection.)

Beehive Coke Ovens Carpentertown, 1968 and 2019. Amanda Kuhns Hollis is pictured below with family members in front of the remains of the dormant beehive coke ovens in Carpentertown in 1968. Despite the continued deterioration of the ovens over time, the integrity of the design of the ovens is still visible today. (Above, author's collection; below, courtesy of the Aaron Hollis Jr. private collection.)

United States Steel Corporation's Clairton Plant (above) Clairton Works, Carnegie-Illinois Steel Mill on the Monongahela River at Night (below) . Eventually, the industry turned to by-product ovens, which reduce the toxic emissions associated with beehive ovens by capturing valuable gases and chemicals to sell for other uses. These ovens were often built near steel mills, and beehive ovens became obsolete. Today, ovens produce coke at Clairton Plant, a division of US Steel in Clairton. (Above, courtesy of United States Steel Corporation; below, courtesy of Rivers of Steel National Heritage Area.)

Reed Brothers/Uniondale Ovens in Dunbar, c. 1950; Refurbished Beehive Coke Ovens on Sheepskin Trail, Dunbar, 2017. The "Connellsville Coke Region Building our Nation's Might" sign seen below states that "at its peak, the region housed 38,000 ovens," but new technologies led to the local coke industry's decline in the 1920s. The oven on the left below is the same one pictured above. (Both, courtesy of Dunbar Historical Society Archives.)

Drawing Coke (above). The Dunbar Historical Society coke oven seen below is outside of the historical society on Bridge Street. According to the society it was built from H.C. Frick's plans and "it is the only safe, user-friendly beehive coke oven in Fayette County today." (Above, courtesy of H.C.F. Negatives, Coal and Coke Heritage Center at Penn State Fayette, the Eberly Campus; below, courtesy of Dunbar Historical Society Archives.)

CHAPTER 3

Organized Labor

Revere Tent City, 1922 Strike. In 1922, miners from the Revere Mine joined the national coal strike. Miners who were evicted by the company from their company-owned homes lived in tents. Production demands, working conditions, wages, and child labor were important issues for the coal, coke, and steel industries. Laborers fought for improved working conditions and higher wages, while management attempted to balance the demands of the new era of industrialization with industry and workplace issues. Labor unions formed and workers went on strike, which sometimes led to violent confrontations. Their efforts, and regulations like the National Labor Relations Act, helped lead to improved wages, safer working conditions, and child labor laws. (Courtesy of Patch Work Voices Community Collection, Coal and Coke Heritage Center at Penn State Fayette, the Eberly Campus.)

The Strike at Homestead. In 1892, Carnegie Steel Corporation's Homestead Mill was the site of a significant event for both labor and management. According to "The Battle of Homestead: A Defining Moment in American Labor History," the collective bargaining agreement between labor and management was to expire at the end of June. When negotiations broke down, plant manager H.C. Frick decided to close the mill and locked the union out. Workers posted men outside to prevent people from entering, to protect from strikebreakers. The pump house, seen here, became central to attempts to stop the Pinkerton guards hired by the company. (Above, author's collection; below, courtesy of Rivers of Steel National Heritage Area.)

Battle of Homestead, 1892 and 2018. Frick hired the Pinkerton Detective Agency to help reopen the mill. On July 6, the Pinkerton barges arrived, and the strike turned violent. Ten people were killed (three Pinkertons and seven workers) and dozens were injured before the Pinkertons surrendered. The Pennsylvania National Guard was sent to regain control of the mill. The plant reopened with non-union workers. The strike eventually collapsed, and unions in the industry were not seen again until the 1930s. The pump house, which was central to the confrontation, is now a trailhead of the Great Allegheny Passage. (Above, courtesy of Rivers of Steel National Heritage Area; below, author's collection.)

Graham Avenue in Windber, 1920s and 2019. The one- and two-story brick buildings in the background on the left side of Graham Avenue are easily identified in both images. Today, there is a historical marker "Windber Strike of 1922-23" in front of the buildings between the two street lights. The sign highlights the labor struggle of the miners and how the "Windber-area Berwind White workers joined a national strike by the United Mine Workers of America in April 1922 for improved wages and working conditions, civil liberties, and recognition." (Above, courtesy of Eureka Coal Heritage Foundation; below, author's collection.)

Morewood Mine and Coke Works in Operation at the Turn of the Century. Workers in Pennsylvania were part of the efforts to obtain higher wages and better working conditions in the late 1800s. H.C. Frick Coke Company's Morewood Coke Ovens near Scottdale were the scene of one such strike. The Morewood Massacre sign on Route 981 identifies the area as being near the site where strikers marched. According to the marker, the strikers were part of 16,000 workers striking for higher wages in the coke regions. The workers marched on Morewood at the end of March and damaged the facility. (Above, author's collection; below, courtesy of West Overton Village and Museum Archives.)

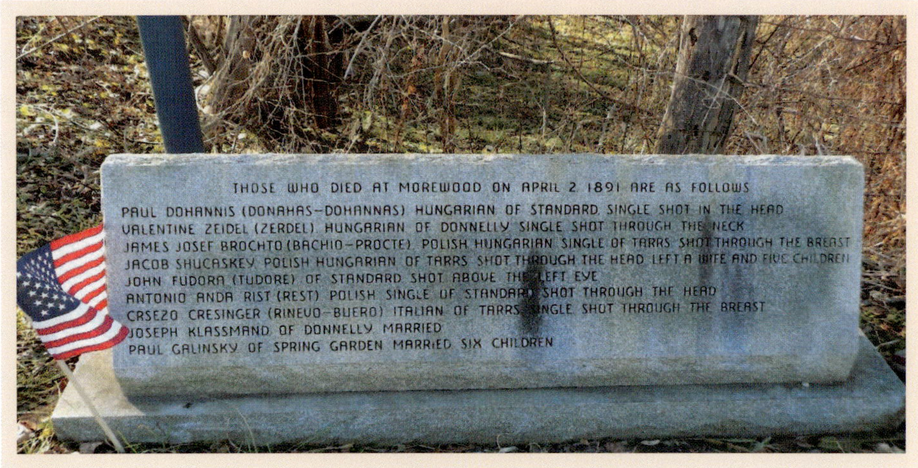

Morewood Mine Complex, August 10, 1934, and 2018. On April 2, 1891, sheriff's deputies killed seven strikers, and two more died later. By late May, the strike had collapsed. Thousands attended the funeral for the victims, and today, the memorial stone near the marker is a reminder of the events at Morewood in 1891. (Above, author's collection; below, courtesy of Patch Work Voices Community Collection, Coal and Coke Heritage Center at Penn State Fayette, the Eberly Campus.)

CHAPTER 4

COMMUNITIES

H.C. FRICK BASEBALL TEAM, 1911 CHAMPS. Many coal and coke operations were in rural areas with limited housing and limited public transportation, so the companies built "patch" or company towns near their operations and rented homes to workers. The towns were nearly self-sufficient. The company store was owned and operated by the company and sold food, clothes, hardware, and other necessities. In some stores, workers could deduct purchases directly from their pay. People tended to live close to others of similar backgrounds, which provided a sense of familiarity for those living in a new community and, in many cases, a new country. Churches, ethnic and social clubs, and sports teams developed in the towns, which created a strong sense of community as people of diverse ethnicities and backgrounds shared customs, foods, and in some cases, religious beliefs. All these things contributed to the unique culture of the region. This chapter includes images from Allegheny, Fayette, Somerset, Washington, and Westmoreland Counties. (Courtesy of Patch Work Voices Community Collection, Coal and Coke Heritage Center at Penn State Fayette, the Eberly Campus.)

BRERETON STREET 1938: VIEW WEST ON BRERETON STREET FROM 3200 BLOCK WITH IMMACULATE HEART OF MARY R.C. CHURCH. Construction of the Immaculate Heart of Mary Catholic Church seen in the distance was finished in 1905. According to *Let's Learn from the Past: Polish Hill*, by the late 1800s, large numbers of Polish immigrants settled in the area known as Polish Hill in Pittsburgh. Neighbors could share customs, foods, and in many cases, religion. (Above, courtesy of Pittsburgh City Photographer Collection, 1901–2002, AIS 1971.05, Archive and Special Collections, University of Pittsburgh Library System; below, courtesy of Pittsburgh History & Landmarks Foundation.)

Pike Street (Smallman Street). These images look similar, with the building-lined street leading to the church. The area in Pittsburgh, along the Allegheny River just outside of downtown known as "the Strip District," has changed over the years. In the early 1900s, wholesale businesses moved into the area. Then some of the businesses opened retail stores. Today, the district is abuzz with sidewalk vendors, produce stands, meat and fish markets, coffee shops, and ethnic stores. The area is a popular destination for residents and tourists. (Above, courtesy of John Viglione; below, courtesy of Pittsburgh City Photographer Collection, 1901–2002, AIS1971.05, Archive & Special Collections, University of Pittsburgh Library System.)

Coal and Coke Arch Looking North, Intersection of Crawford Avenue and Pittsburgh Street, Connellsville. The arch seen below was built by H.C. Frick Coke Co. for the town's centennial celebration, which began on August 14 and ended on August 18, 1906. It cost $10,000 and was lit up at night. The arch was removed the week of August 20. Today, there is no evidence of the arch, but the building to the right is the same in both images. The company was owned by Henry Clay "H.C." Frick, who was born in Westmoreland County. The image above is from 2018. (Above, author's collection; below, courtesy of Thomas E. Rusnack.)

Coal and Coke Arch, Connellsville, Looking East. The trolley and tracks are gone, but the buildings on either side of the arch are easily recognized in both images. H.C. Frick, seeing the potential of the bituminous coal region to respond to the growing steel industry's need for coke, got into the coke business. His business dominated the Connellsville coke region in Fayette and Westmoreland Counties, and he eventually entered into a partnership with Andrew Carnegie (see chapter six). The image below is from 2019. (Above, courtesy of Thomas E. Rusnack; below, courtesy of John Viglione.)

Graham Avenue, c. 1920s and 2018. The town of Windber is in the northern section of Somerset County. The name is an anagram for the company that built the town, Berwind-White Coal Mining Company. According to *A Legacy of Coal*, the idea was to build a large independent town surrounded by smaller dependent mining communities that were built near mining operations. In addition to the company's office building, the town also had amenities like a bank, privately owned shops, and a theater. The one-story building and the theater across the street are easily recognized in both images. The theater's name, "Arcadia," is clearly visible on the sign. The theater remains an active part of the community today. (Above, courtesy of Dan Hiteshew; below, author's collection.)

Eureka Department Store and Bank, c. 1906 and 2019. According to *A Legacy of Coal*, the company's subsidiary, Eureka Supply Company Ltd., administered all company-owned stores. There were over 30 stores, including several in Windber. The main Eureka store on Fifteenth Street in Windber (below, second building from the corner) expanded into the bank building around 1916. An English Tudor style was added to the exterior (above). Today, the landmark building remains on the corner. Portions of the building are occupied by local businesses. (Above, author's collection; below, courtesy of Dan Hiteshew.)

Office Building, 1930s and 2019. According to *A Legacy of Coal*, the two-story building seen here was financed by the Berwind-White Coal Mining Company in 1913. Today, there is a brick wall to the left of the building that is not seen below. (Above, author's collection; below, courtesy of Eureka Coal Heritage Foundation.)

The Windber Electric Building. According to "A Legacy of Coal" it was Built in the 1920s. A portion of the Berwind-White Coal Mining Company building can be seen in the background. The street view of the building and the company's office building are remarkably similar today. (Above, courtesy of Eureka Coal Heritage Foundation; below, author's collection).

Marianna, Summer of 1913 and 2018. The Pittsburg & Buffalo Company, owners of the Marianna Mine, built what was considered a model mining town at the time, according to *A Bicentennial History of West Bethlehem Township and Marianna Borough 1776–1976*. Many homes were constructed of yellow brick. The yellow bricks were also used to build other structures in town. The homes had conveniences that were not always typical of mining towns, like bathrooms and sewage. The yellow bricks are clearly visible below. (Above, courtesy of Marianna Community Public Library; below, author's collection.)

Daisytown Company Store and Gas Station. The company store provided essentials for miners and their families and was also a central meeting place in many towns. In the 1950s, a department store, grocery store, and butcher shop were all part of the company store (seen here at right) in Daisytown. There were gas pumps next to the store, and the store also made home deliveries. Miners who worked at Vesta No. 4 had a charge account at the store, and purchases would be deducted from their pay. (Above, author's collection; below, courtesy of Ellen R. Keder.)

Christmas at the Standard Union Supply Company Store in 1912 and 2018. One of the largest company store chains in Western Pennsylvania was the Union Supply Company. Incorporated in the late 1890s, the chain served the coal and coke plants of the H.C. Frick Coke Company. Note that the "Standard" sign on top of the building (the name of the mine it serviced) is now gone, and there is an addition to the main entrance in front. (Above, author's collection; below, courtesy of the Mount Pleasant Historical Society.)

Coke Arch, Mount Pleasant, 1910 and 2019. Henry C. Frick was a large sponsor of a weeklong Old Home Week event in Mount Pleasant the week of July 4 in 1910. The arch was built by H.C. Frick Coke Company, according to a July 5, 1910, article in the *Connellsville Daily Courier*. The arch went over Main Street near the corner of Diamond Street. The support was coal, and coke was used for the rest of the arch. (Above, courtesy of Rick Meason; below, author's collection.)

Coke Arch at Night, 1910. Similar to the arch in Connellsville, the Mount Pleasant arch lit up and illuminated the area at night. The initials "H.C.F.C. Co." were at the crest of the arch. The arch was taken down shortly after the event. Rides, booths, and entertainment filled the streets during the weeklong festivities. Today, a statue of a World War I US Army doughboy is in the middle of the intersection. It was dedicated in memory of those who served. (Above, courtesy of Rick Meason; below, author's collection.)

CHAPTER 5

DISASTERS

MARIANNA MINE DISASTER, NOVEMBER 28, 1908. The Marianna Mine was the site of an explosion that killed 154 in November 1908 (see page 78). There were different—and in some cases, multiple—causes for accidents at that time. Open-flame lighting and limited industry regulations were some of the factors cited. The push for safer working conditions led to changes in the industry and the formation of regulatory agencies. Advances in the technology of mining methods and equipment and improved regulations have greatly improved mine safety. (Courtesy of Patch Work Voices Community Collection, Coal and Coke Heritage Center at Penn State Fayette, the Eberly Campus.)

Hill Farm Mine Explosion. On June 16, 1890, an explosion at the Hill Farm Mine, owned by the Dunbar Furnace Company, killed 31 miners. Smoke can be seen at the mine scene at center below. According to the plaque outside the Dunbar Historical Society, the explosion was one of the most deadly in the history of the coke region. Only two victims were recovered at the time of the disaster. The other 29 were not recovered until 1892. The 2019 photograph shows the general site of the mine today. (Above, author's collection; below, courtesy of Frederick T. Gretton Photographs, Detre Library & Archives, Heinz History Center.)

Robena Mine H.C. Frick Coke Company. On December 6, 1962, around 1:00 p.m., methane gas exploded at the mine, killing 37 miners according to a 2012 Triblive article about the accident. The mine is closed, but there is a memorial on Route 21 in Greene County in memory of the men who were injured and those who lost their lives at the Robena Mine. (Right, courtesy of US Steel Corporation Duquesne Works Industrial Relations Department Records, 1904–1980, AIS.1987.03, Archives and Special Collections, University of Pittsburgh; below, courtesy of Jonathan Tajc.)

MARIANNA MINE DISASTER, NOVEMBER 28, 1908, AND 2018. Just before 11:00 a.m. on November 28, an explosion rocked the Marianna Mines, a Pittsburg-Buffalo Company operation, killing 154. The accident was a little over a month after Pres. Theodore Roosevelt invited mining experts from Europe to Marianna as part of an inspection of US coal industries (see page 30). Although the exact location of the image above is not identified, the remains of coke ovens from the operation can be seen below (at right, behind power lines). (Above, courtesy of Patch Work Voices Community Collection, Coal and Coke Heritage Center at Penn State Fayette, the Eberly Campus; below, courtesy of John Viglione.)

The Mammoth Mine. On January 27, 1891, an explosion at Mammoth No. 1 killed 109 miners. There was also a slope entry, Mammoth No. 2, at the mine, and although there were some injuries from the explosion, the Mammoth No. 2 miners escaped according to 2002 Triblive article. Through preservation efforts, the Mammoth No. 2 portal entry, behind the Mount Pleasant Township Municipal Authority building seen in chapter one, has been restored. The below picture of the Mammoth Mine is from the *Engineering and Mining Journal* of February 7, 1891. (Above, author's collection; below, courtesy of [E&MJ] *Engineering and Mining Journal*.)

Group of Coke Drawers at Mammoth Works (below). Today, near the Mount Pleasant Township Municipal Authority Building on page 39 there is a memorial to remember the disaster. The memorial, above, is near the restored portal and lists the names of the victims of the disaster. (Above, author's collection; below, courtesy of Hagley Museum and Library.)

Darr Mine Disaster. On December 19, 1907, an explosion at the Darr Mine killed 239 in the worst mine disaster in Pennsylvania history. Today, along the Great Allegheny Passage near Van Meter, there are two Darr Mine markers. The "Darr Mine Disaster" sign is left near a memorial stone and the "Cable Car" sign is in the foreground right. The cable/ferry car was used to transport miners who lived across the river in Jacobs Creek. The above picture of Jacobs Creek looks west across the Youghiogheny River. The mine would have been located on the opposite side of the river near the left flume of smoke. (Above, courtesy of Donald Lancaster; below, author's collection.)

RESCUE PARTY AT THE ENTRANCE OF THE DARR MINE. The Darr Mine disaster occurred in December 1907. The *Report of the Department of Mines of Pennsylvania, Part II Bituminous, 1907* stated that "Persons in the vicinity of the mine described the explosion as an awful rumbling followed by a loud report and a concussion that shook the nearby buildings and was felt within a radius of several miles." The mine entrance would have been on the hillside behind the sign and memorial along the trail pictured below. (Above, courtesy of Donald Lancaster; below, author's collection.)

Rescue Party Outside the Darr Mine. Official numbers say 239 were killed. St. Nicholas Day was also part of the Darr Mine story. According to *The Report of the Department of Mines*, around 400 men worked at the mine, but that day many miners stayed home to celebrate a festival of the Greek Catholic Church. The lives of those who stayed home were spared that day. The monument along the trail (above) placed by the Belle Vernon Rotary Cub describes the disaster as "one of the worst mining disasters in the nation." (Above, author's collection; below, courtesy of Donald Lancaster.)

Unidentified Dead from the Darr Mine Explosion. Today, the common grave of 71 of the victims can be found at the Olive Branch Cemetery. The monument reads, "To the Memory of the Martyrs of the Darr Mine Disaster, Dec. 19, 1907. Erected by the Hungarian-American Federation 1909." (Above, author's collection; below, courtesy of Donald Lancaster.)

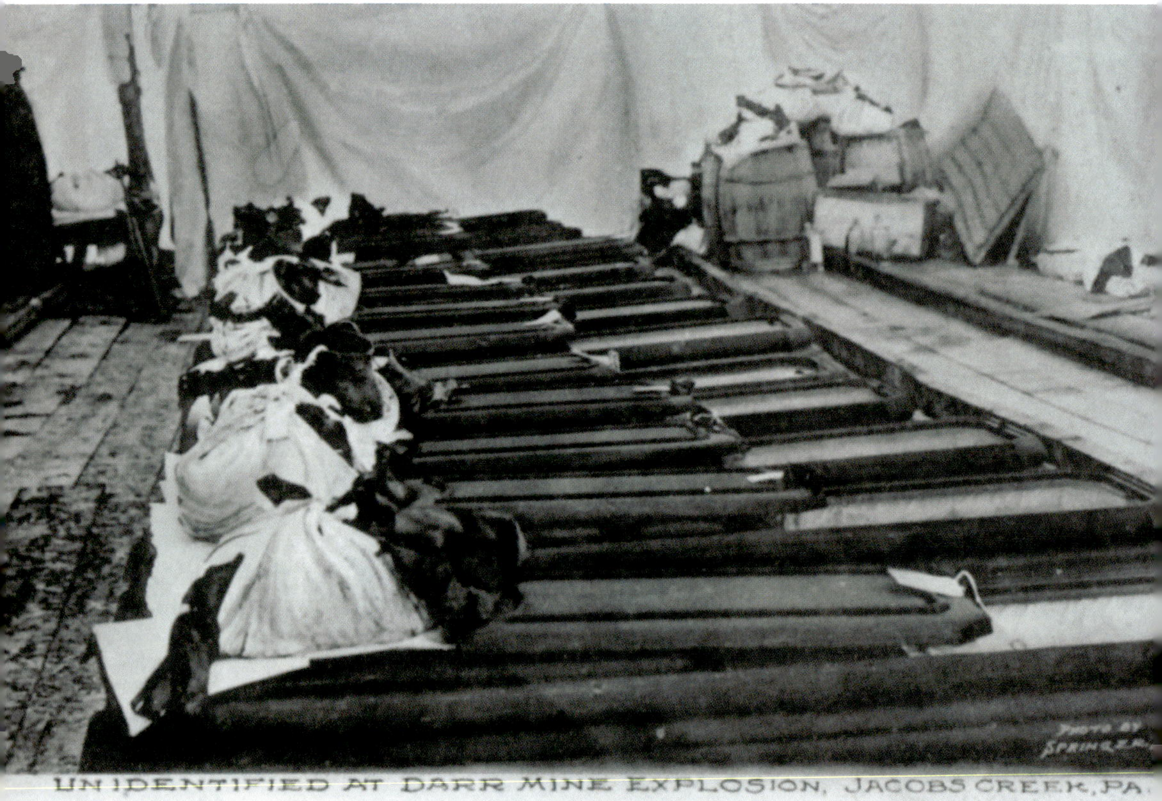

CHAPTER 6

Industry Leaders

Union Arcade Construction. Those who saw the potential for industrial development in the area took risks and ventured into uncharted areas as the region and the industrialization that was spreading through other parts of the country merged. Leaders of the industries came from various backgrounds and achieved local, and in some cases, international success. Fortunes were made and lost as the coal, coke, and steel industries developed in the region. Although these leaders are long gone, the influence of some of them in the region is still a part of society today. The above building, which was known as the Union Arcade, was built by Henry Clay Frick opened in 1917. (Courtesy of Union Arcade Building Photograph Collection, 1915–1916, AIS.2005.09, Archives & Special Collections, University of Pittsburgh Library System.)

Fraternal Order of Eagles/Peter Campbell Building, Scottdale. According to the Scottdale Historical Society, the building is where Frick had his first office located on the second floor, "the F.O.E. did a circa 1949 remodel of the building applying stucco to the brick façade, post war art deco influenced entry door [below]." Seeing the potential in the coke region, Frick opened a coke plant in the early 1870s. He continued to build the business and eventually dominated the coke industry, producing coke used by the growing iron and steel businesses in Pittsburgh. (Both, courtesy of Scottdale Historical Society Inc.)

OFFICES OF THE H.C. FRICK COKE COMPANY, SCOTTDALE. According to the Scottdale Historical Society, "As Frick's business grew, he moved a block further south on South Broadway. There Frick constructed the Victorian office building with third-floor dormers in circa 1880. The business continued to flourish, and the larger three-story building was constructed on the corner of Broadway and Walnut Street [left]." Today, the building to the right is a vacant lot. Frick began providing coke to the Carnegie mills and eventually entered a partnership with Carnegie. (Both, author's collection.)

Clayton, the Historic Home of the Henry Clay Frick Family (above) Clayton Exterior, 1901 (below). H.C. Frick and his family moved to Pittsburgh in the 1880s. They lived at Clayton House until they moved to New York in 1905, where he died in 1919. Today, the public can tour the Clayton House, the Frick Art Museum, the gardens, and other facilities on the grounds of the Frick Pittsburgh. (Above, courtesy of the Frick Pittsburgh; below, courtesy of the Clayton Album, 1901, Frick Family Albums and Scrapbooks, the Frick Collection/Frick Art Reference Library Archives.)

Carnegie Library, Connellsville. Andrew Carnegie immigrated with his family from Scotland to Pittsburgh in the late 1840s. He achieved success in various businesses before entering the steel industry, where he built one of the largest steel companies in the world, which he eventually sold. He also devoted his efforts to philanthropy. His efforts included the establishment of non-profit organizations, museums (like the Carnegie museums in Pittsburgh) and other institutions like the Carnegie Free Library in Connellsville. The library opened in 1903 and is listed in the National Register of Historic Places. The past and present images are remarkably similar. (Both, author's collection.)

Built by Philanthropist Sarah Cochran in Memory of Her Husband Philip. The Philip G. Cochran Memorial Methodist Church is as beautiful today as it was when it was built. Philip was the son of James Cochran, founder of the Washington Coal and Coke Company mentioned in chapter one. Philip took over running the successful family business after his father's death. He lived with his wife, Sarah, and their son in Dawson, where the church is located. (Above, courtesy of the Fayette County Historical Society; below, author's collection.)

J.V. Thompson's Oak Hill Estate. Designed by famed Chicago architect D.H. Burnham, the Josiah Van (J.V.) Kirk Thompson mansion was constructed in the early 1900s. Now owned by the Sisters of the Order of St. Basil the Great, it is now Mount Saint Macrina House Of Prayer. The stately pillars and details of the mansion still stand, much as they did when it was first built. J.V. Thompson was a prominent banker who made a fortune buying and selling coal fields. He lost his fortune and the mansion. He was allowed to remain in the mansion until he died in 1933. (Above, photograph by author; below, courtesy of Fayette County Historical Society.)

J.V. Thompson's Skyscraper, the First National Bank, 1902 and 2018. J.V. Thompson became president of First National Bank after his father died. The bank became an important part of the community. As seen at right, J.V. began construction of the First National Bank Building in Uniontown; the architect was Daniel Burnham. The photograph appears to show the beginning of the construction of the entrance of the building. (Above, author's collection; right, courtesy of the Fayette County Historical Society.)

J.V. Thompson's First National Bank, 1902 and 2018. The building was designed by internationally renowned architect Daniel Burnham, whose work included the Flatiron Building in New York City. Today, the building is known as the Fayette Building, and it, like Oak Hill, is a reminder of the era of coal and coke. (Left, courtesy of Fayette County Historical Society; below, author's collection.)

Bibliography

"Andrew Carnegie Collection—Biography." diva.library.cmu.edu/webapp/carnegie/bio.html.

"Banning No. 1: Processing Coal for the Steel Mills. Southwestern Pennsylvania Heritage Preservation Commission, Allegheny Trail Alliance.

Basinger, Rachel R. "Approximately 275 Witness Mammoth Mine Memorial Unveiling." archive.triblive.com/news/approximately-275-witness-mammoth-mine-memorial-unveiling. October 5, 2002.

"The Battle of Homestead: A Defining Moment in American Labor History." Rivers of Steel. riversofsteel.com/attractions/pump-house/the-battle-of-homestead/.

Brestensky, Dennis F., Evelyn A. Hovanec, and Albert N. Skomra. *Patch/Work Voices: the Culture and Lore of a Mining People*. Pittsburgh, PA: University of Pittsburgh Press, 1991.

Carnegie Free Library of Connellsville. www.carnegiefreelib.org.

"Clearfield Bituminous Coal Corporation land tract records and property maps, circa 1885–2009." www.libraries.psu.edu/findingaids/1493.htm. Special Collections Library, Pennsylvania State University.

Coal and Coke Resource Analysis Western Pennsylvania, Northern West Virginia. Denver, CO: US Department of the Interior, National Park Service, Denver Service Center, 1992.

"Coke Making." fayette.psu.edu/visit/coalandcoke/museum. The Coal and Coke Heritage Center, Penn State Fayette, the Eberly Campus.

"The Connellsville Coke Region." The *Engineering Magazine*. October 1900–March 1901: 17-40.

"Connellsville Coke Region Building Our Nation's Might." The Progress Funds Trail Town Program.

Connellsville Daily Courier, July 5, 1910

Edmunds, William E. "Coal In Pennsylvania." www.docs.dcnr.pa.gov/cs/groups/public/documents/document/dcnr_014594.pdf. Pennsylvania Geological Survey. 2002.

"First Mining of Pittsburgh Coal" historical marker. explorepahistory.com/hmarker.php?markerId=1-A-2C5.

Gillespie, Charles R. "Work of Rescue Goes Slowly Forward Under Guard of State Police—Still Enough Gas to Blow up Entire Hill." *The Pittsburg Press*, December 20, 1907.

"Henry Clay Frick." www.frick.org/about/history/henry_clay_frick.

Maher, Regis M. *Patches of History: The 1920's & 1930's Heyday of Fayette County Coal & Coke in Pennsylvania*. Self-published, 1999.

"The Mammoth Mine Disaster." *Engineering and Mining Journal* Vol. 51, 1891.

"Morewood Massacre" historical marker. explorepahistory.com/hmarker.php?markerId=1-A-234.

Mulrooney, Margaret M. *A Legacy of Coal: the Coal Company Towns of Southwestern Pennsylvania*. Washington, DC: Historic American Buildings Survey/Historic American Engineering Record, National Park Service, 1989.

Napsha, Joe. "Greene County Mine Tragedy Recalled, 50 Years Later." archive.triblive.com/news/pennsylvania/greene-county-mine-tragedy-recalled-50-years-later/

National Register of Historic Places registration form for Oake Hill Estate. Denver, CO: US Department of the Interior, National Park Service, 1999.

Pitz, Marylynne. "December 1907 was Most Disastrous Month in U.S. Coal History." *Pittsburgh Post-Gazette*. November 28, 2007.

Potisek, Lillian, and Singadine Muchant. *A Bicentennial History of West Bethlehem Township and Marianna Borough, 1776–1976*. Marianna, PA: 1976.

Report of the Department of Mines of Pennsylvania. Part II: Bituminous. Harrisburg, PA: Harrisburg Publishing Co., 1908.

Schano, Ned. "Let's Learn From the Past: Polish Hill." *Pittsburgh Post-Gazette*. August 18, 2011.

Storey, Jerry. "Mammoth Mine Portal to History Opens in Mt. Pleasant Township." *Tribune-Review*. September 29, 2002.

Templeton, David. "A Past, but No Future: Suspension Bridge Coming Down." *Pittsburgh Post-Gazette*. February 29, 2004.

"A Walk Through Time: Pennsylvania Coal Culture." Special Collections and University Archives, University Libraries, Indiana University of Pennsylvania. libweb1.library.iup.edu/depts/speccol/All%20Finding%20Aids/R&Pcatalog.pdf. 2009

"Windber Strike of 1922–1923" historical marker. explorepahistory.com/hmarker.php?markerId=1-A-2CF.